Immunology A to Z

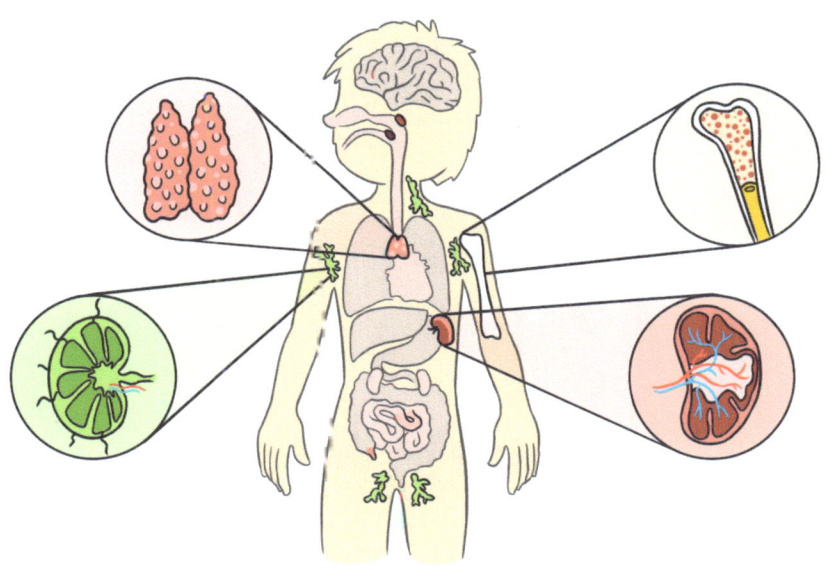

Written by Ella Maughan, PhD

Illustrated by Martina Pepiciello, MSc

A

Antigens
(an-ti-jens)

Proteins found on the surface of most cells in your body and on germs. Your **immune cells** know which antigens are on your body's cells ("self") or germs ("non-self"). When they see non-self-antigens, they form an **immune response**.

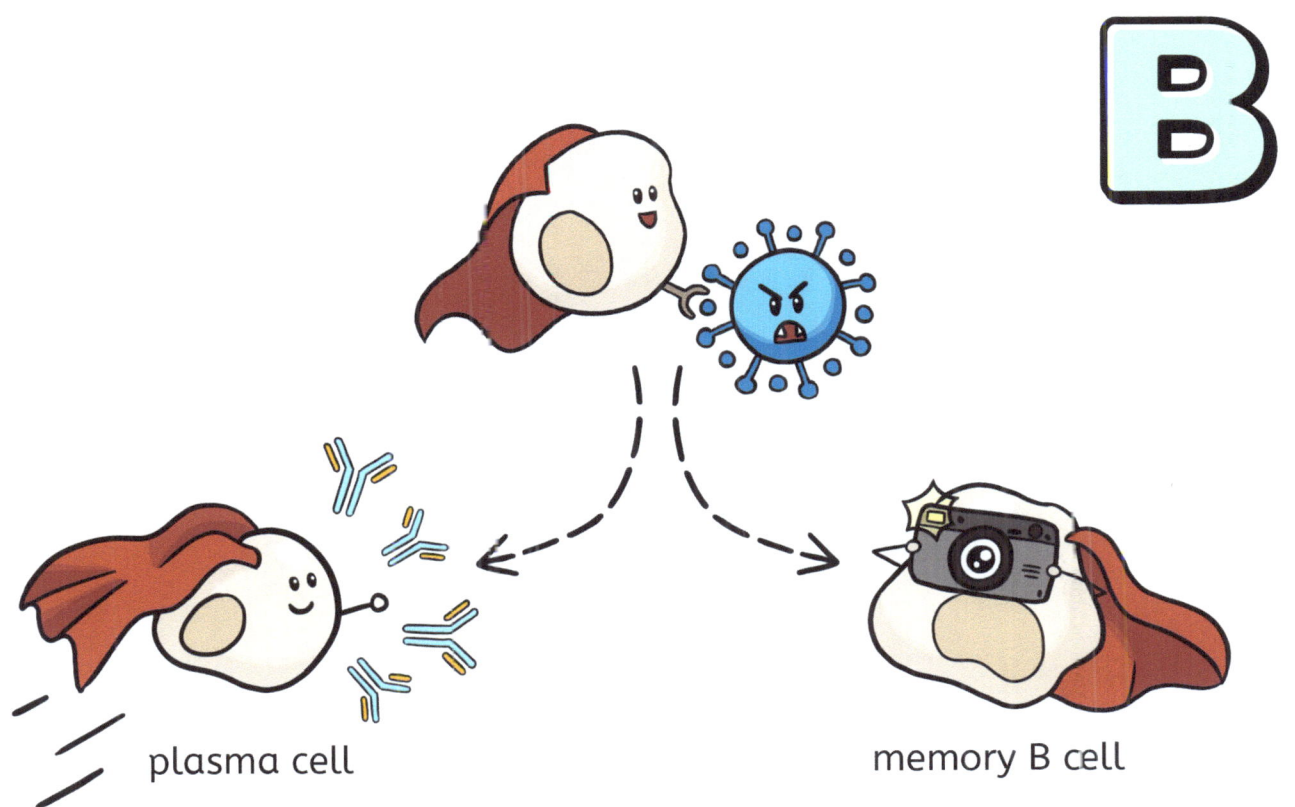

B cells
(*bee selz*)

White blood cells that fight germs and abnormal cells. When B cells spot something that doesn't belong, they turn into plasma cells that make **antibodies** or memory B cells that help your body remember the threat.

C

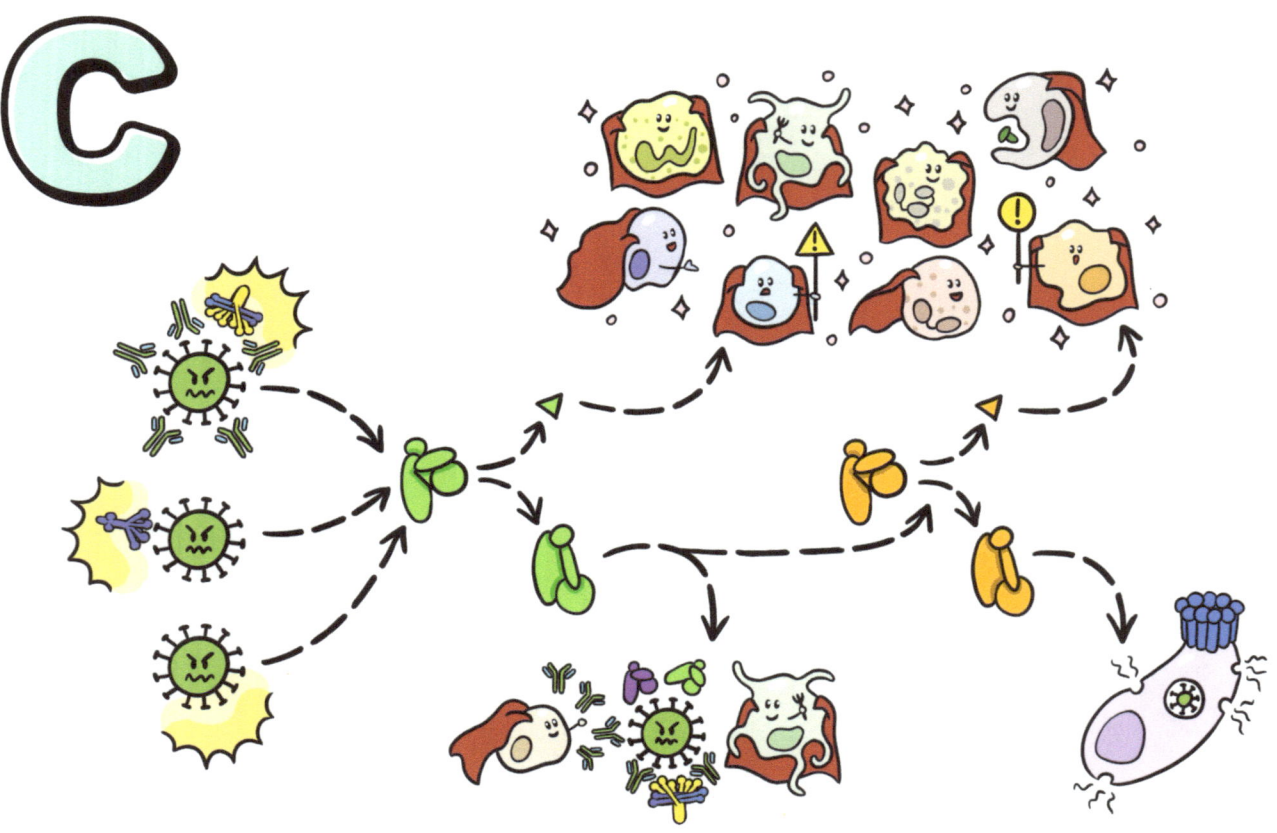

Complement system
(KOM-pleh-ment SIS-tuhm)

A group of proteins that help your immune system. It activates through different pathways, triggering chain reactions that can fight germs, control **inflammation**, clean up damage, and heal injuries.

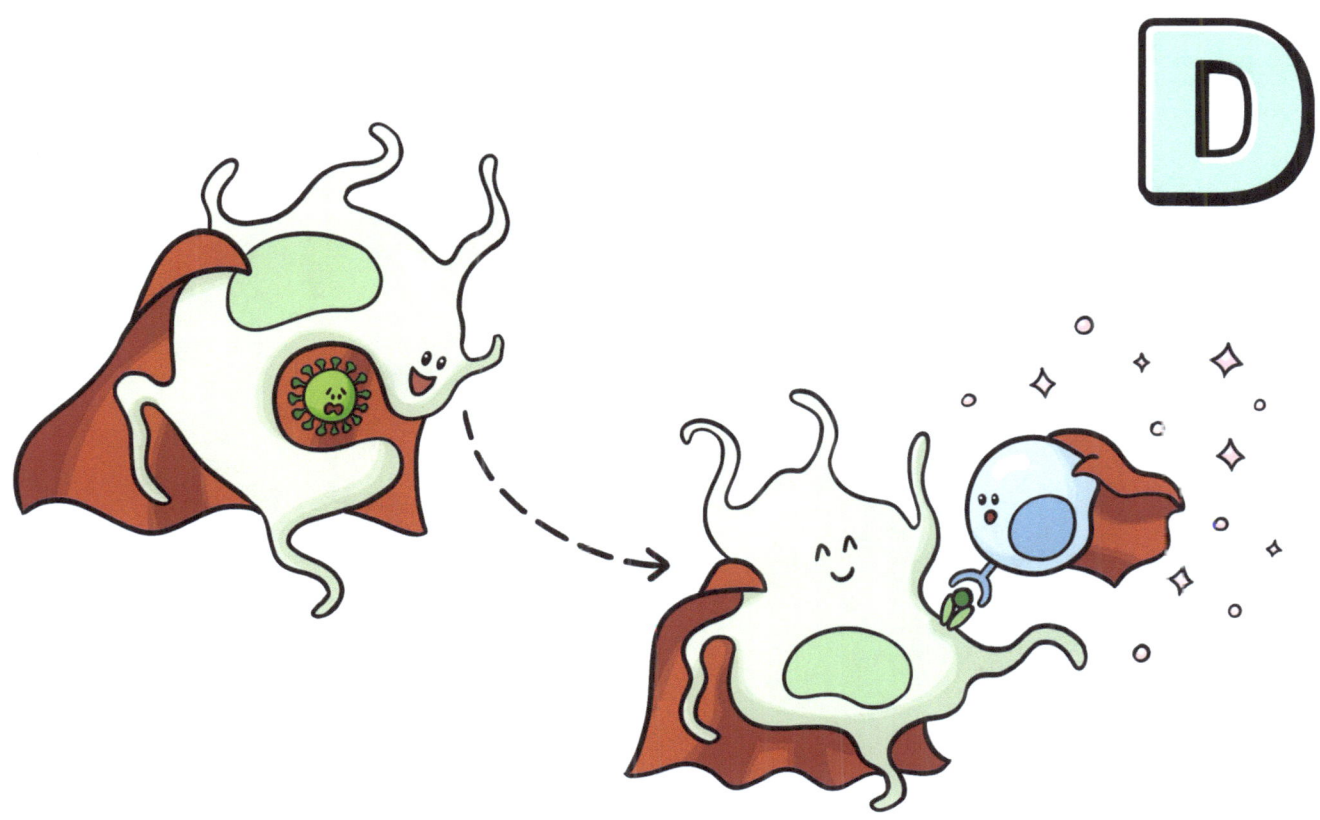

Dendritic cells

(den-DRIH-tik selz)

A type of immune cell called a **phagocyte.**
They gobble up germs and damaged or abnormal cells.
After breaking them into pieces, they show the antigens to
helper T cells to activate an immune response.

E

Eosinophils

(ee-oh-SIN-uh-fills)

White blood cells that fight germs (like parasites and fungi) and trap **toxins** and **venoms**. They also help heal wounds and keep your immune system balanced. But if eosinophils are too active, they can cause **allergies** and **asthma**.

Fever

(FEE-vur)

Raised body temperature, usually caused by **infection**. During an infection, immune cells send messengers to your brain to turn up the heat, which helps your body fight germs faster.

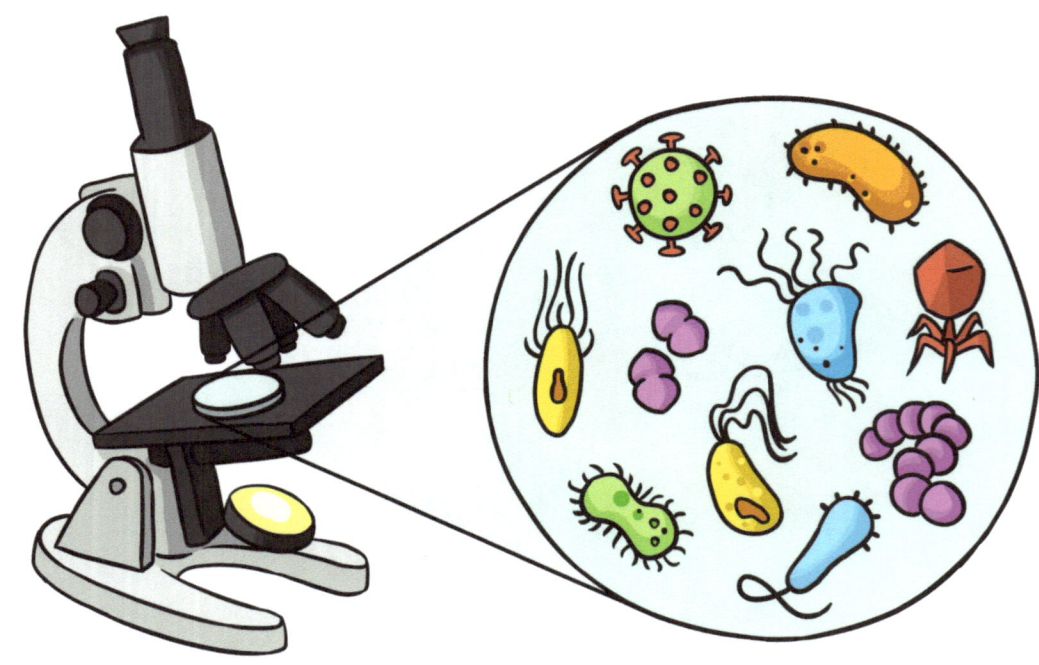

Germs

(JERMS)

Tiny organisms—bacteria, viruses, fungi, and protozoa—that can make you sick. Germs spread through coughs, sneezes, saliva (spit), blood, touch, and dirty food or water.

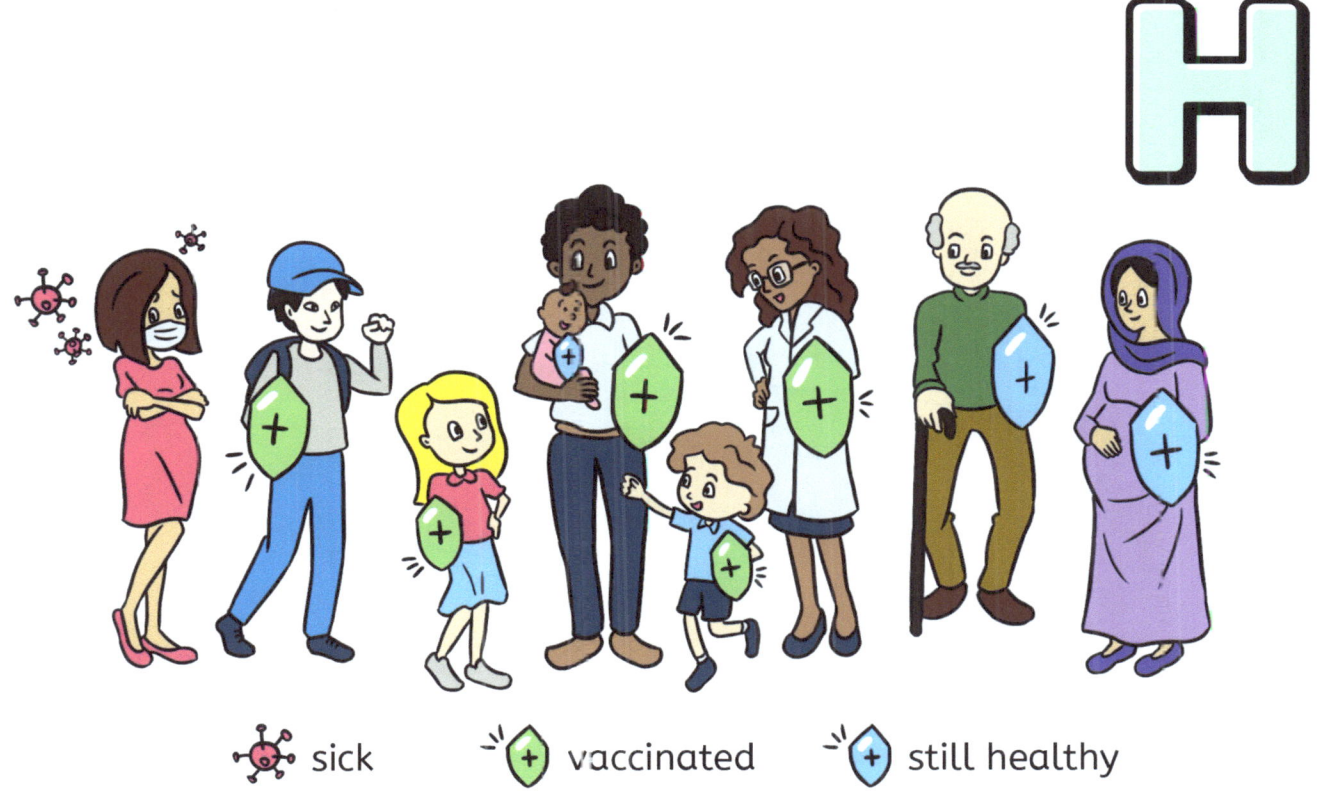

Herd immunity

(HURD ih-MYOO-nih-tee)

When enough people in a community are **immune** to a disease to slow the spread of infection. Herd immunity protects everyone, including newborn babies and those who can't get vaccines.

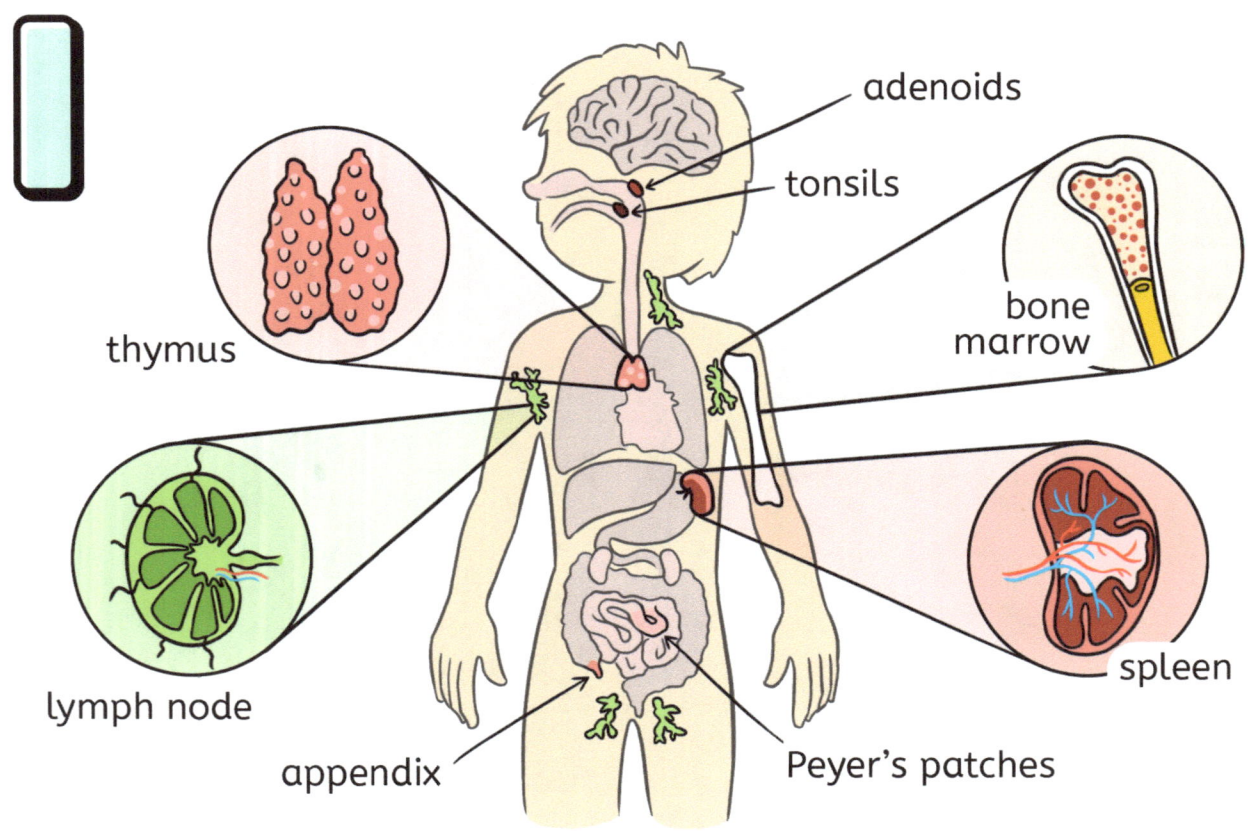

Immune system

(ih-MYOON SIS-tum)

Your body's defense team! It includes special immune cells, **tissues**, **organs**, and **chemical messengers** that work together to fight germs, heal injuries, and keep you healthy.

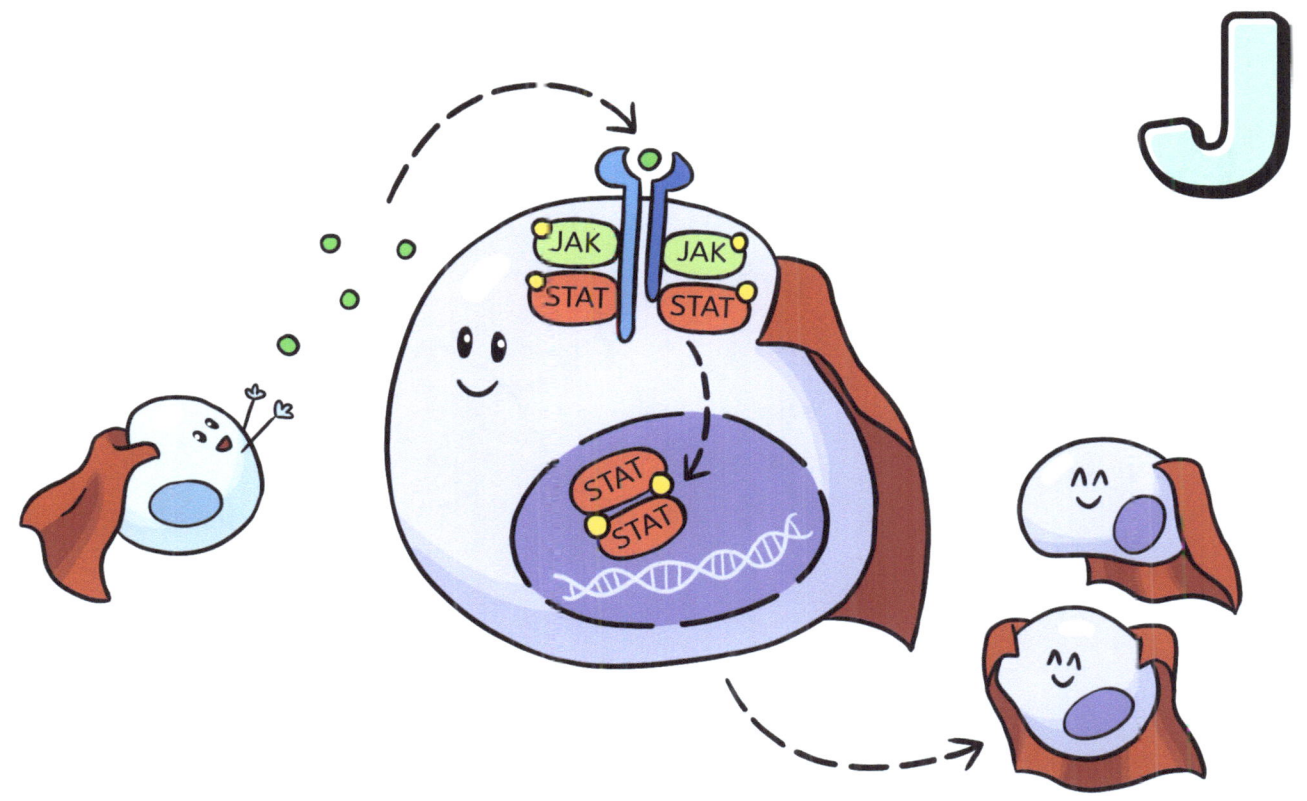

JAK-STAT signaling
(JACK-STAT SIG-nuh-ling)

How cells communicate with each other to control your immune response. When a messenger reaches a cell, helpers called **JAK** and **STAT** carry it inside and tell the cell what to do, like making more immune cells or turning on a response.

K

Killer T cells

(*KILL-er tee selz*)

Immune cells that search for and destroy infected and abnormal cells. They also release messengers that tell nearby cells to stay alert, help other immune cells fight, and stop viruses from multiplying before they get inside cells.

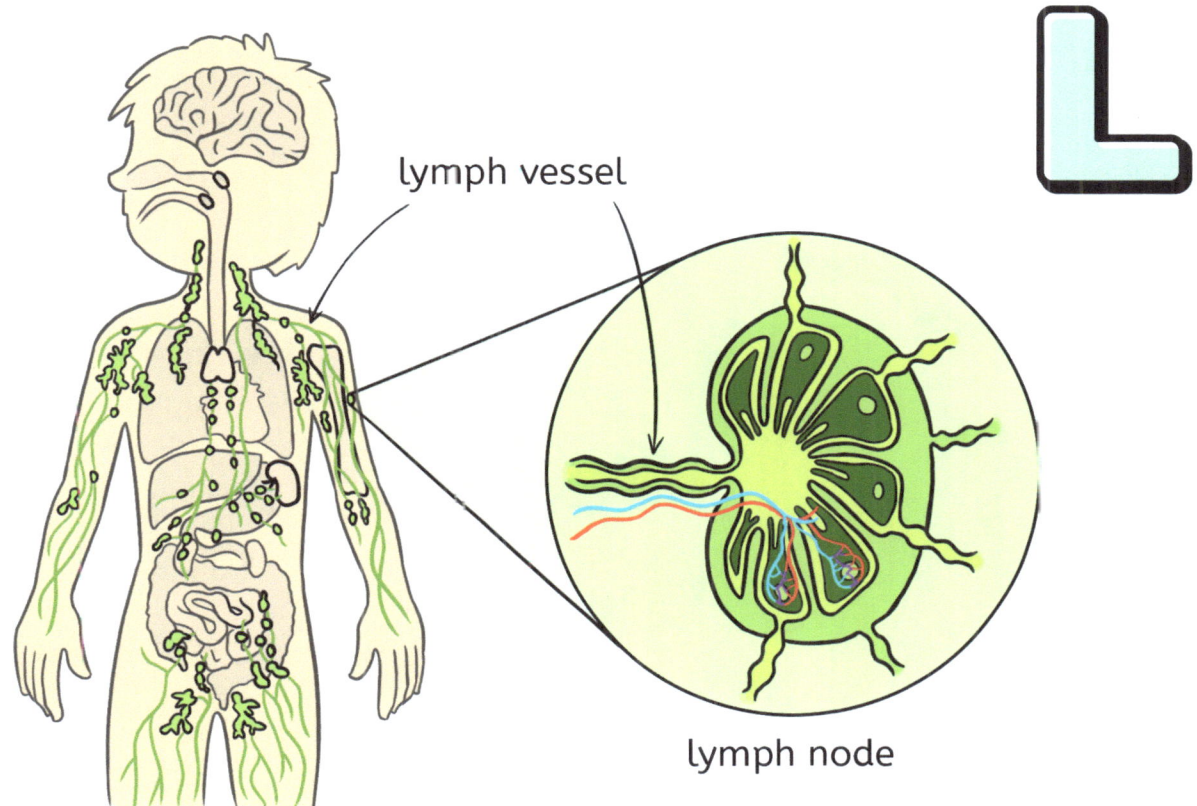

Lymph nodes

(*LIMF nohds*)

Small, bean-shaped organs that clean your body's **lymph** fluid. They trap germs, waste, and unhealthy or abnormal cells. Inside, **lymphocytes** fight off any threats they find.

M

Mast cells
(MAST selz)

Immune cells that live in tissues and protect against harmful invaders. When they sense danger, they release messengers that call other immune cells to help fight threats and start healing. When mast cells overreact, **allergic reactions** occur.

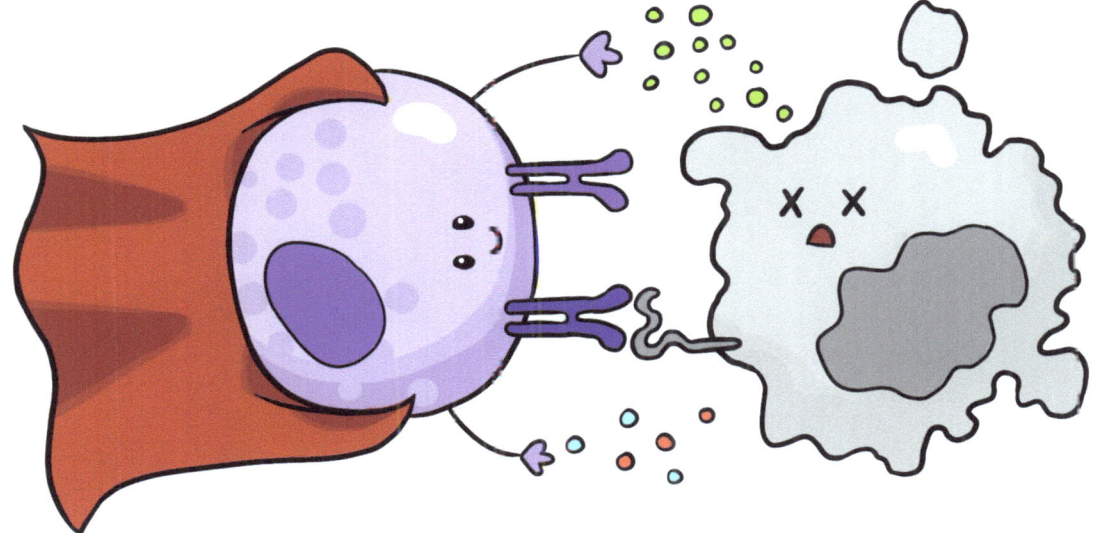

Natural killer (NK) cells

(NACH-er-uhl KILL-er selz)

Fast-acting immune cells that patrol your body for infected or abnormal cells. NK cells destroy infected cells before germs can spread. They also remove unhealthy cells before they cause more damage.

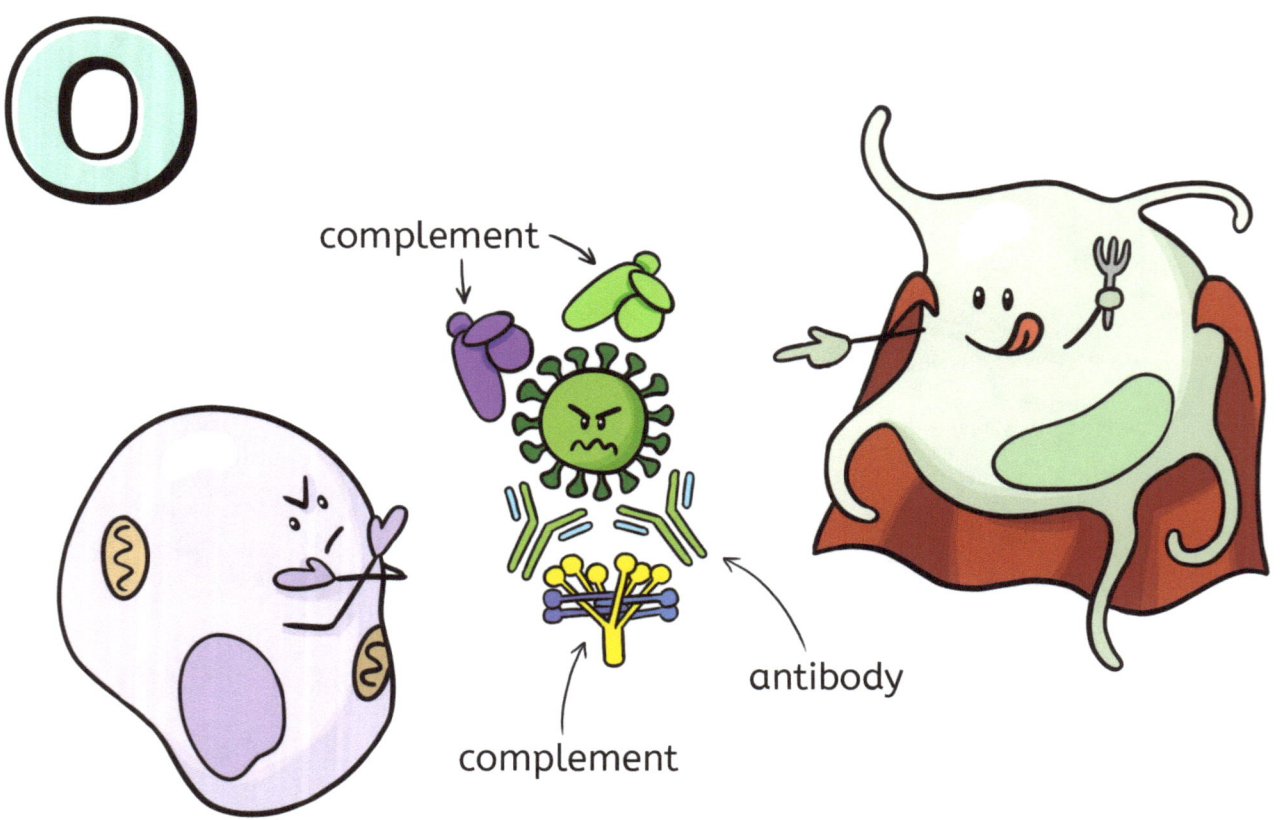

Opsonization
(op-suh-ny-ZAY-shun)

When antibodies and complement proteins stick to germs and damaged cells. This process tags them so immune cells can quickly find and destroy them. It also blocks germs from sneaking into healthy cells.

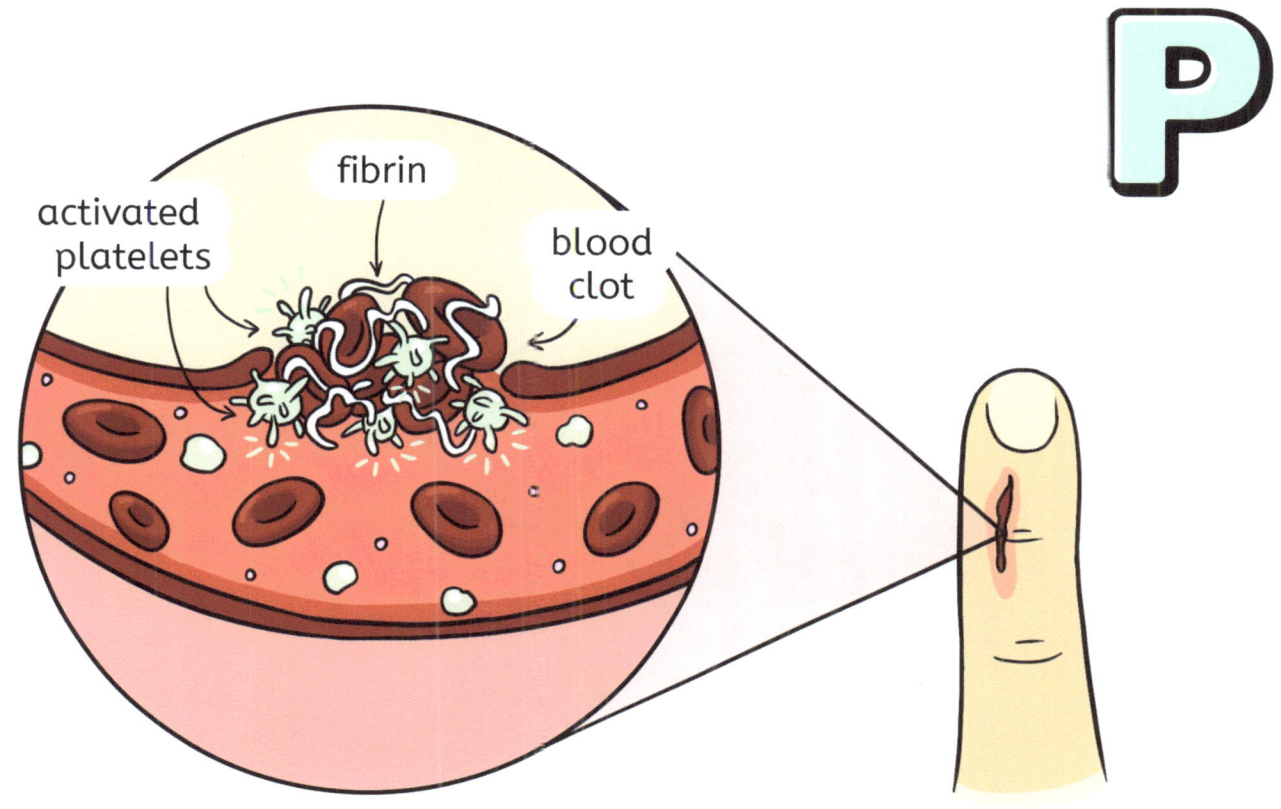

Platelets

(*playt-lits*)

Tiny cell pieces that help stop bleeding. When a blood vessel is injured, platelets activate and stick to the wound. **Fibrin** (*fye-brin*) then forms a sticky net over them, creating a blood clot that seals the injury and helps it heal.

Q

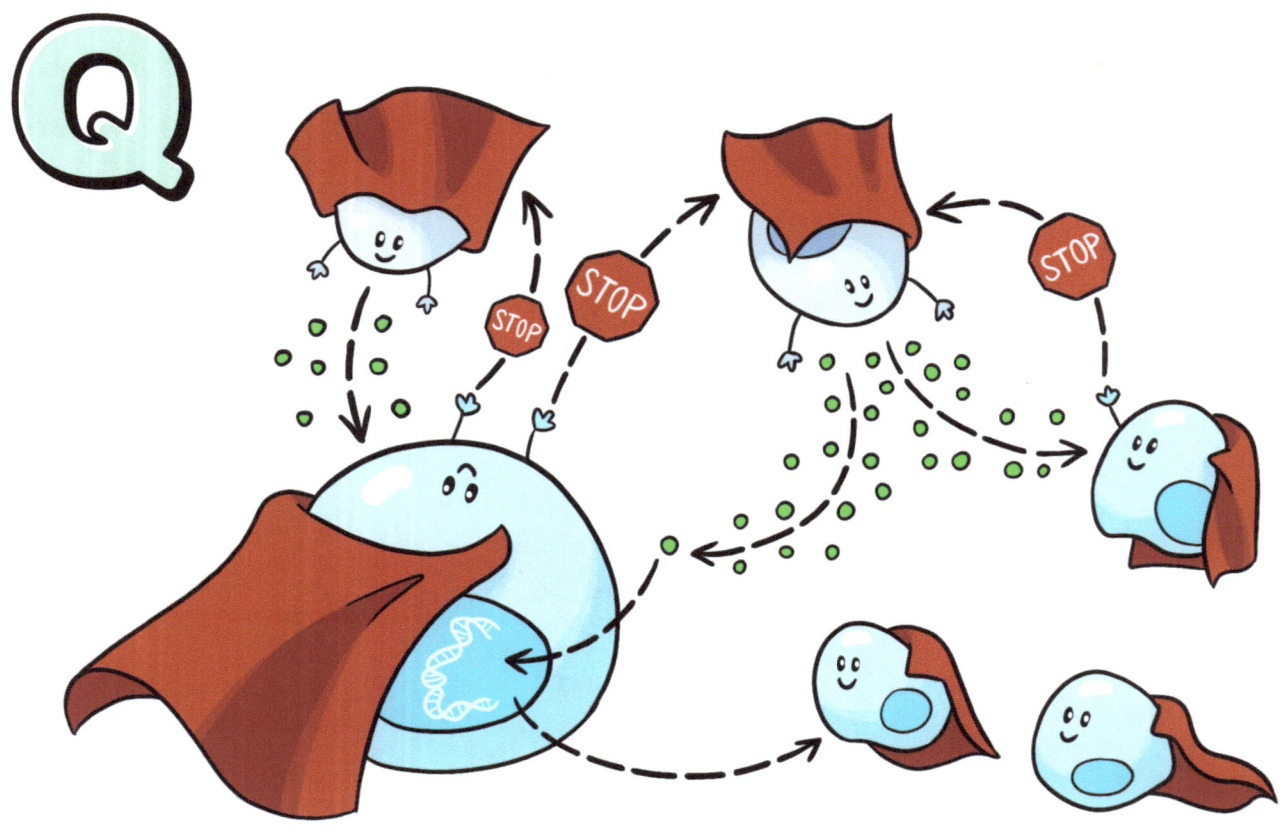

Quorum sensing

(KWOR-uhm SEN-sing)

How immune cells communicate using chemical messengers. It helps them know when to team up and fight hard and when to slow down, keeping your immune system balanced.

Regulatory T cells

(REG-yoo-luh-TOR-ee tee selz)

Immune cells that help keep the immune system in check. They stop it from overreacting or attacking your own body, preventing **allergies** and **autoimmune diseases**.

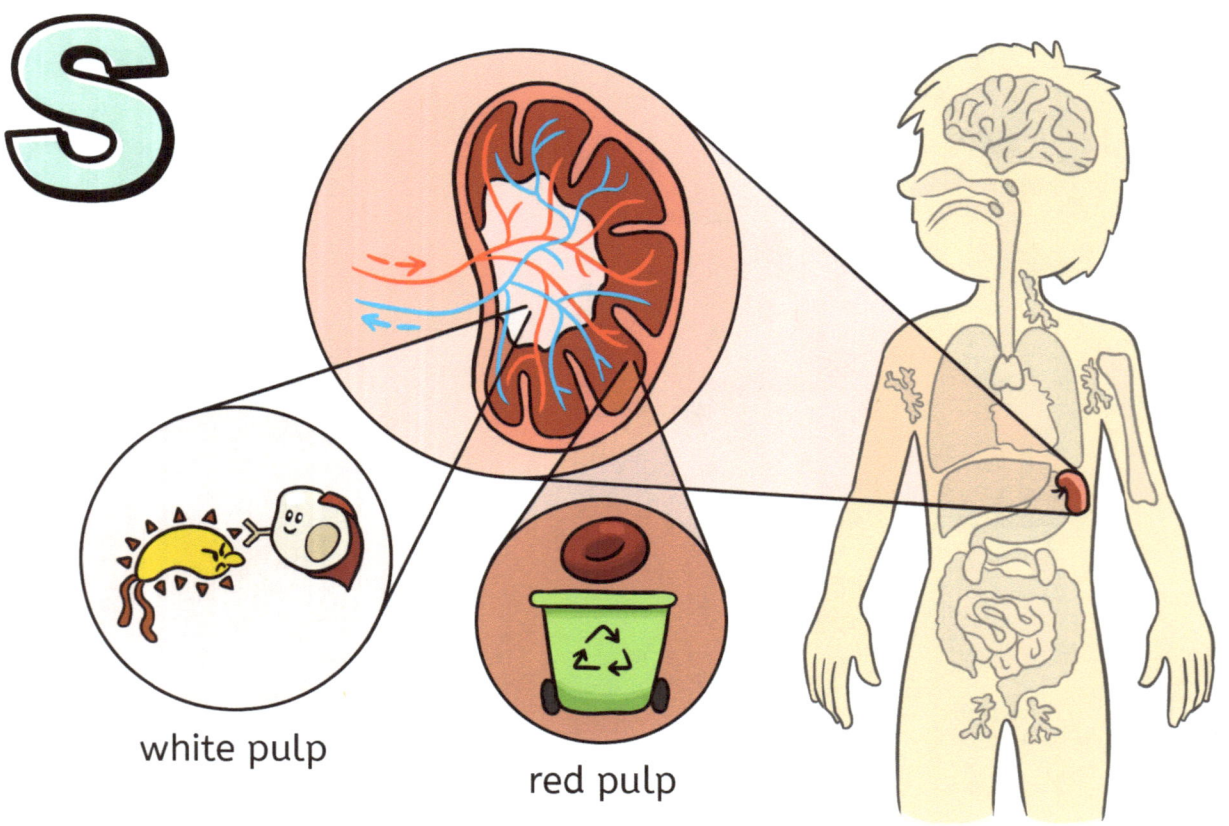

Spleen

(*spleen*)

A fist-sized organ that sits under your ribs.
It filters and recycles old and damaged red blood cells.
It also stores white blood cells that check your blood for germs and abnormal cells to help fight illness.

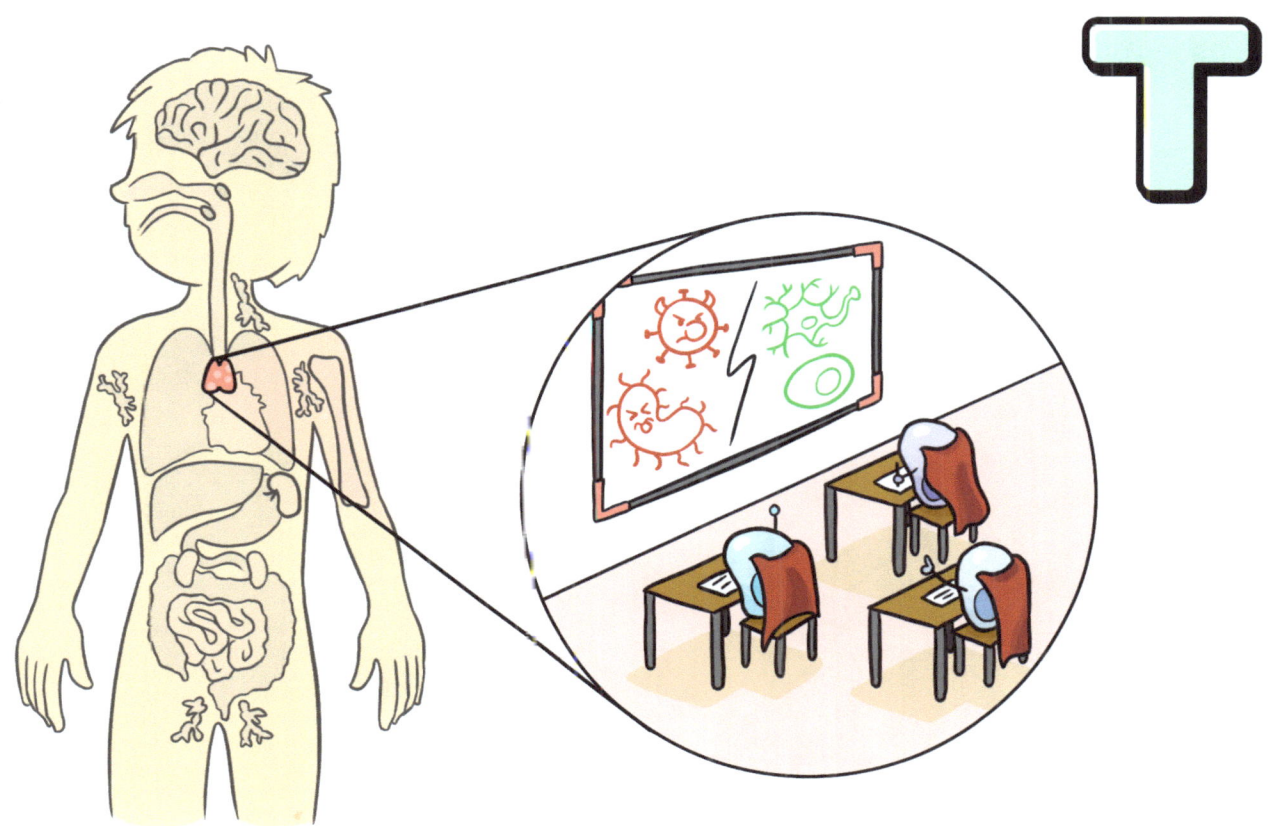

Thymus

(THY-mus)

An organ in your chest where T cells learn the difference between your own cells and harmful invaders. Here, they grow into different types, like regulatory, killer, and helper T cells.

U

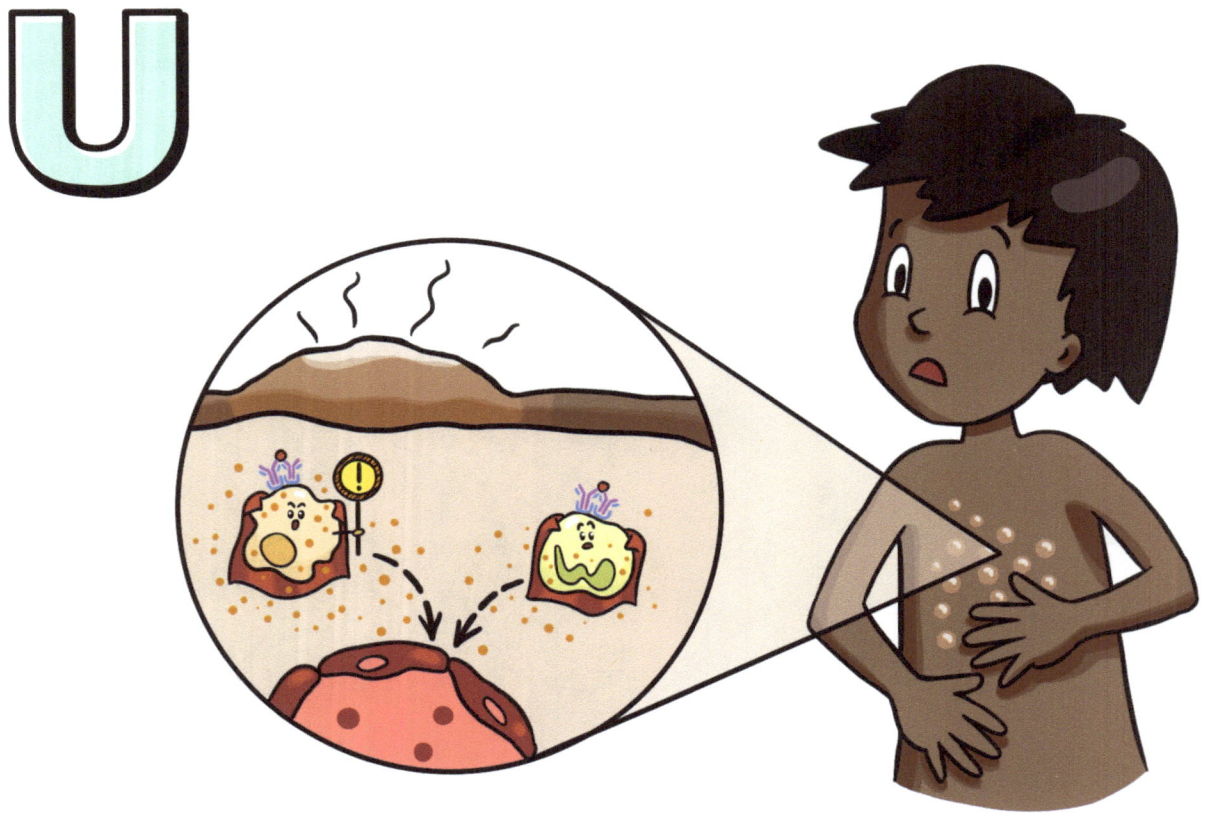

Urticaria

(ur-tuh-CARE-ee-uh)

The medical word for hives—raised, itchy patches on the skin. Hives appear when immune cells release histamine and other messengers due to allergies, infections, physical triggers (like heat or pressure), or autoimmune diseases.

Vaccines

(vack-SEENZ)

Medicines that train your immune system to fight germs. They show it harmless versions of germs so your immune cells can practice fighting, build memories, and quickly respond to real germs if they see them later.

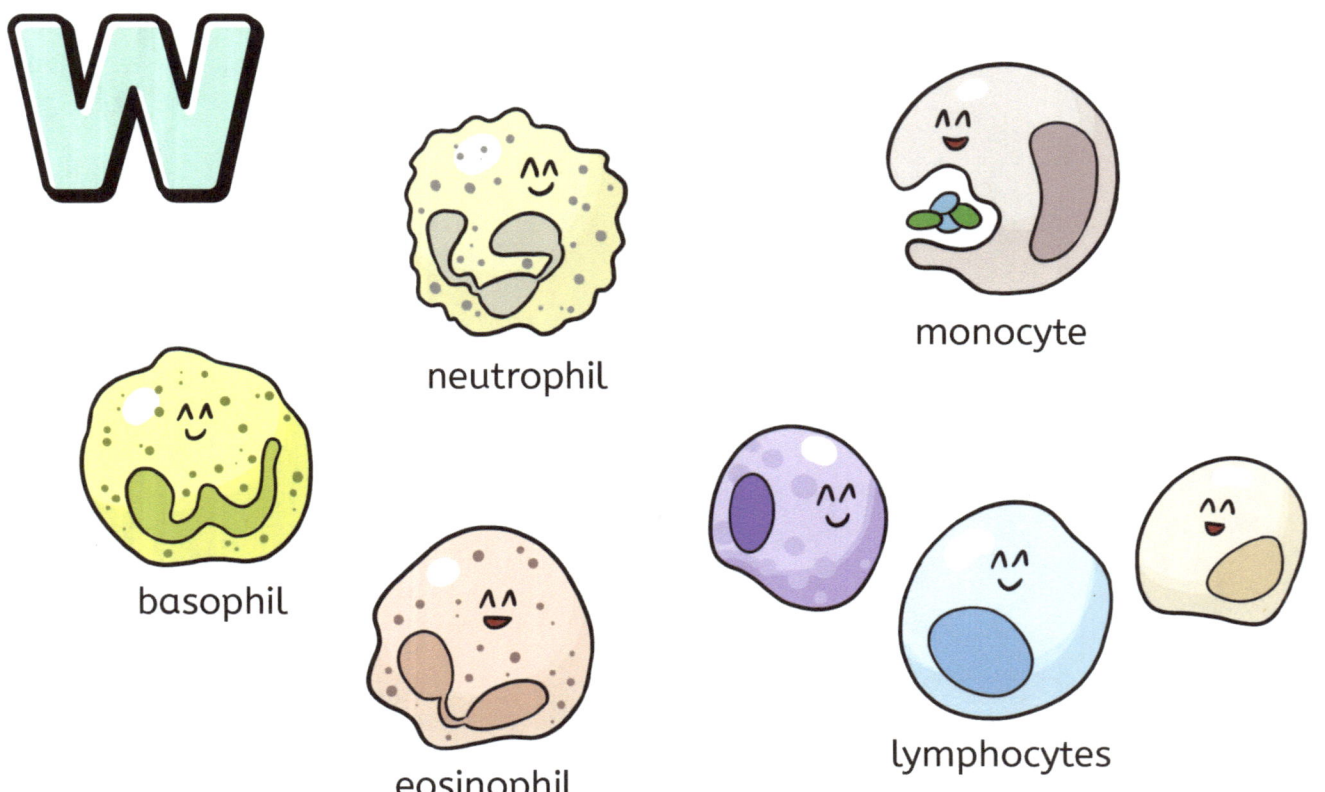

White blood cells

(WYTE blud selz)

Immune cells that are made in your **bone marrow** and travel through your bloodstream. Each type has several roles, from fighting germs to removing abnormal cells, cleaning up damage, or triggering allergic responses.

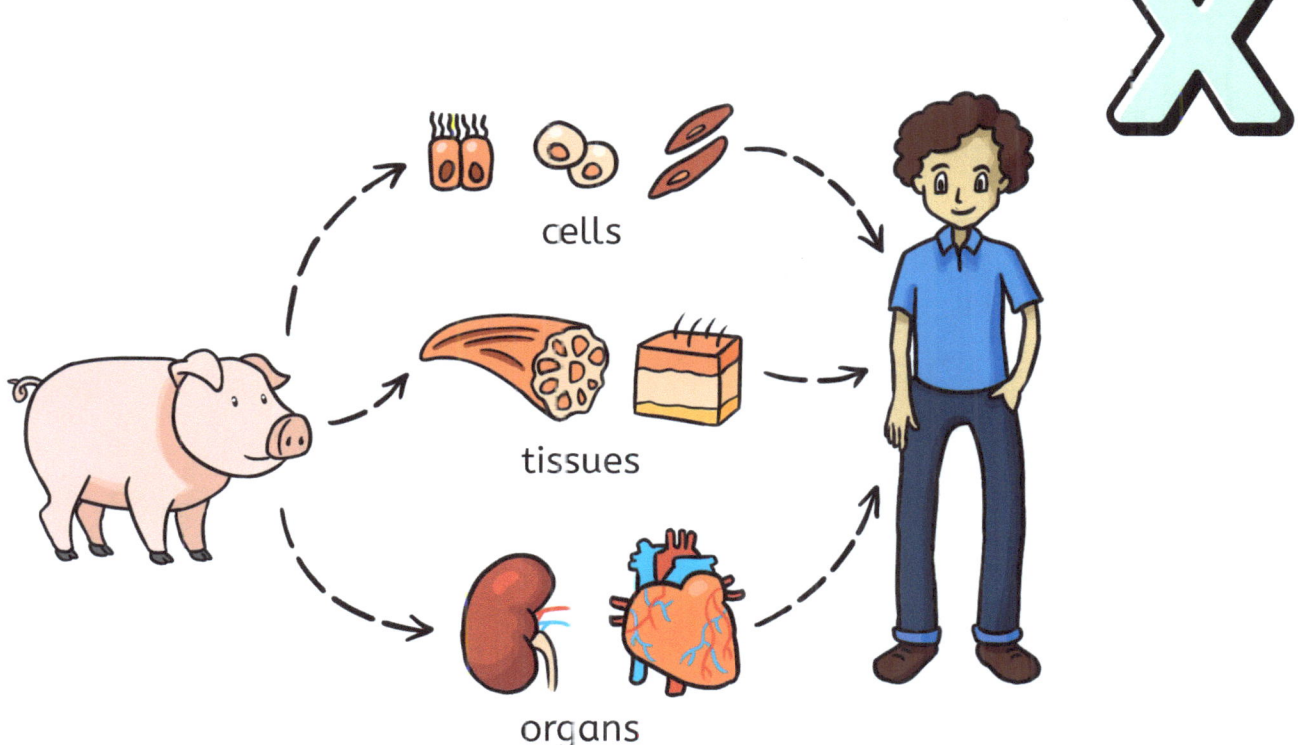

Xenotransplantation

(ZEE-noh-trans-plan-TAY-shuhn)

When living cells, tissues, or organs are moved from one species to another. Doctors and scientists are working to make these **transplants** safe by modifying them to prevent infections and rejection by the immune system.

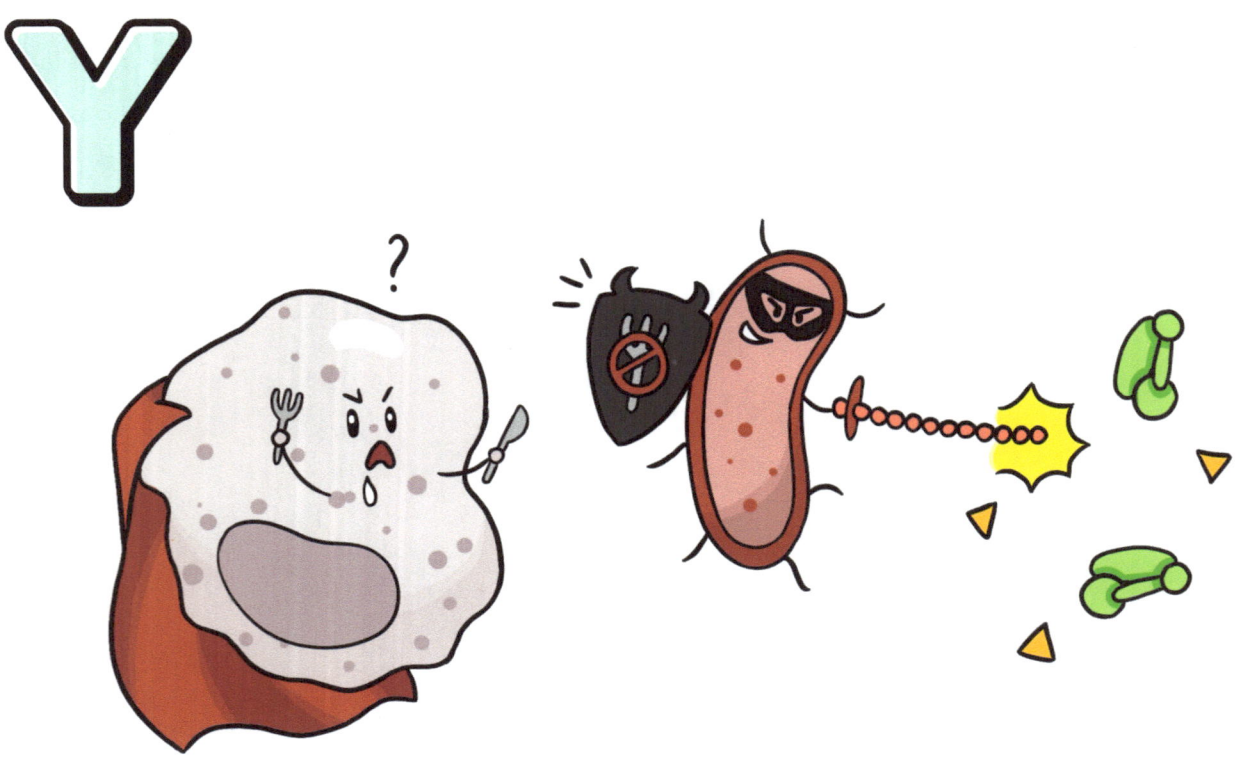

Yersinia pestis

(yer-SIN-ee-uh PES-tis)

Bacteria that cause **plague**. They escape the immune system by blocking complement proteins, stopping immune signals, and stealing nutrients, before spreading quickly through the body.

Zoonosis

(*zoh-uh-NOH-sis*)

A disease that spreads between animals and humans. Germs can spread through touch, bites, food, water, or the environment. Scientists study how the immune system responds to these germs to prevent **outbreaks.**

GLOSSARY

Adenoids: Tissue at the back of the nose that traps germs. They contain immune cells that fight infections. Fun fact: your adenoids shrink as you get older!

Allergies: Abnormal reactions of the immune system to usually harmless **allergens**, like pollen, dust, or peanuts. These cause **allergic reactions** like itching, sneezing, or trouble breathing.

Asthma: A condition where the lung airways swell and narrow, leading to chest tightness, coughing, wheezing, and trouble breathing. It can be triggered by allergens, exercise, infections, or cold air.

Antibodies: Y-shaped proteins made by plasma B cells after an infection or vaccination. Antibodies block germs from infecting cells and signal other immune cells to attack.

Appendix: A small worm-shaped organ attached to your large intestine. It may act as a "bank" for good bacteria in the gut and serve as a site for immune training.

Autoimmune diseases: Conditions where the immune system mistakenly attacks the body's own healthy cells, tissues, and organs, as if they were harmful invaders.

Basophil: A type of white blood cell that defends your body against allergens and infections, and helps prevent blood clots forming in damaged tissues.

Bone marrow: Soft, spongy tissue inside your bones that makes red blood cells, white blood cells, and platelets.

Cells: The basic units of all living things. Some germs are made of only one cell. Your body is made up of trillions of cells!

Chemical messengers: Small proteins that tell immune cells where to go and how to act, to guide your immune response and control inflammation. Immune cells make different types of chemicals, including signaling molecules like cytokines and histamine, and cytotoxic molecules like granzymes and perforin.

Fibrin: Sticky protein threads that form a net to stop bleeding and seal a wound after you get hurt.

Helper T cell: T cells that coordinate and control your immune response by activating other immune cells like phagocytes, B cells, and killer T cells.

Immune: Protected from infection because of past illness, vaccination, or passive immunity (receiving antibodies made by another human or animal).

Immune cells: Cells that work together to create an immune response, such as B cells, basophils, dendritic cells, eosinophils, mast cells, monocytes, natural killer cells, neutrophils, and T cells.

Immune response: Your immune system's coordinated reaction to infections and other diseases.

Infection: When germs enter the body, multiply, and cause illness.

Inflammation: Your body's response to infection, injury, or illness. Normally, inflammation helps the body to heal, but it can cause harm if it goes on too long.

JAK (Just Another Kinase): A protein inside cells that carries immune signals, telling the cell what to do next.

Lymph: Fluid that flows through your lymphatic system in lymph vessels. It delivers nutrients, collects damaged cells and germs, and returns extra fluid from tissues to the blood.

Lymphocyte: A type of white blood cell that protects your body from germs and harmful cell changes. The main types of lymphocytes are B cells, T cells, and NK cells.

Monocyte: A type of white blood cell and **phagocyte.** Monocytes patrol your blood and tissues for danger. They turn into macrophages and dendritic cells after reaching infected or injured areas.

Neutrophil: The most common type of white blood cell in your body. These fast-moving cells are often the first responders to infections, gobbling up germs and starting the tissue-repair process.

Organ: A group of tissues that work together to perform a specific function. Your brain, heart, lungs, liver, kidneys, and spleen are all types of organs.

Outbreak: When many people in one place get sick from the same disease at the same time.

Peyer's patches: Small areas of immune tissue in your intestines that watch for germs coming in through food or drink.

Phagocytes (*FAY-go-sites*): White blood cells that eat germs in a process called **phagocytosis** (*FAY-go-sigh-TOH-sis*). Examples include monocytes, macrophages, dendritic cells, and neutrophils.

Plague: A rare but serious disease caused by the zoonotic bacteria *Yersinia pestis*. It can infect the lymph nodes, blood, or lungs.

Proteins: Molecules made up of chains of amino acids. Proteins make up your muscles, organs, and immune system. They are essential for your body to function properly.

STAT (Signal Transducer and Activator of Transcription): A family of proteins that carry messages from outside the cell to the nucleus, and control how the cell responds to the message.

Tissue: A group of cells that work together to perform a specific job in the body. There are four main types: epithelial, connective, muscle, and nervous tissue.

Tonsils: Immune tissues at the back of the throat that trap germs entering through the mouth or nose to protect the body from infection.

Toxins (*TOK-sins*): Poisons made by germs, plants, or animals that can harm the body's cells. Your immune system works to fight off toxins to keep you safe.

Transplant: When doctors replace a sick or damaged organ with a healthy one.

Venoms (*VEN-ums*): Poisonous liquids that some animals, like bees, snakes, spiders, scorpions, and jellyfish, inject into their prey or enemies through a bite or sting.

For my girls.

Publisher's Cataloging-in-Publication Data

Names: Maughan, Ella, author. | Pepiciello, Martina, illustrator.
Title: Immunology A to Z / Ella Maughan, PhD; illustrated by Martina Pepiciello, MSc.
Description: New York, NY: Ella Maughan, PhD, 2025. | Summary: Explore immunology—from Antigens to Zoonosis! Complex science is transformed into clear, engaging snapshots of how our bodies fight germs, heal injuries, and stay healthy.
Identifiers: LCCN: 2025924915 | ISBN: 979-8-9889191-3-1 (hardback) | 979-8-9889191-4-8 (paperback) | 979-8-9889191-5-5 (ebook)
Subjects: LCSH Immunology--Juvenile literature. | Immune system--Juvenile literature. | Immunity--Juvenile literature. | Alphabet. | BISAC JUVENILE NONFICTION / Science & Nature / Biology | JUVENILE NONFICTION / Science & Nature / Anatomy & Physiology | JUVENILE NONFICTION / Health & Daily Living / General
Classification: LCC QR181.8 .M38 2025 | DDC 616.079--dc23

Text copyright © 2025 by Eleanor Maughan.
Illustrations copyright © 2025 by Martina Pepiciello.
Edited by John Briggs.

All rights reserved.

No portion of this book may be reproduced, distributed, or transmitted in any form or by any means, without written permission from the publisher or author, except as permitted by U.S. copyright law.

The information provided in this book is intended for educational and general informational purposes only. It is not intended to provide specific medical guidance or recommendations and is not a substitute for professional medical advice.

Ella Maughan has a PhD in Immunology and Microbial Pathogenesis ("Bad Bugs") and has spent more than 10 years working in medical communications and scientific consulting. She is the author of *Bad Bug Busters: Vaccines vs Germs* and *Immunology A to Z*.

Perfect for curious kids, parents, and educators, her books bring science to life through clear explanations and vibrant illustrations—helping children build confidence, STEM vocabulary, and a lifelong interest in how their amazing bodies work.

Originally from Surrey, England, Ella moved to New York in 2015. She lives on Long Island, NY, with her husband, two daughters who love to do "experiments," and their cat, Pickle.

Visit her online at www.ellamaughan.com.

Martina Pepiciello has been passionate about drawing since she learned to hold a pencil and about science since she learned to write with that pencil. The latter interest drove her to earn a Master's Degree in Theoretical Physics at the University of Bologna, Italy. After finishing her studies, during which she never stopped drawing, she merged her passions and become a scientific illustrator and graphic designer.

Martina's goal is to clearly and colorfully depict scientific concepts that might sound intimidating so that they become more accessible. Lately, she has been focusing on depicting the most pressing scientific and technical challenges of our time, engaging different audiences on subjects like climate change, artificial intelligence, and biosecurity through graphics.

She can be found online at www.martinapepiciello.com.

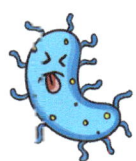